PHYSICAL SCIENCE

PROPERTIES OF MATTER

By Christina Earley

A Stingray Book

SEAHORSE PUBLISHING

Teaching Tips for Caregivers and Teachers:

This Hi-Lo book features high-interest subject matter that will appeal to all readers in intermediate and middle school grades. It may be enjoyed by students reading at or above grade level as well as by those who are looking for age-appropriate themes matched with a less challenging reading level. Hi-Lo books are ideal for ELL readers, too.

Each book appeals to a striving reader's age and maturity level. Opportunities are provided for students to read words they already know while encountering a limited number of new, high-interest vocabulary words. With these supports in place, students will read more fluently while increasing reading comprehension. Use the following suggestions to help students grow as readers.

- Encourage the student to read independently at home.
- Encourage the student to practice reading aloud.
- Encourage activities that require reading.
- Establish a regular reading time.
- Have the student write questions about what they read.

Teaching Tips for Teachers:

Before Reading

- Ask, "What do I know about this topic?"
- Ask, "What do I want to learn about this topic?"

During Reading

- Ask, "What is the author trying to teach me?"
- Ask, "How is this like something I already know?"

After Reading

- Discuss how the text features (headings, index, etc.) help with understanding the topic.
- Ask, "What interesting or fun fact did you learn?"

TABLE OF CONTENTS

WHAT IS MATTER?

Matter is anything that has mass and takes up space.

Properties are used to describe matter.

Some properties can be **observed**.

Tools are used to **measure** other properties.

MASS

Mass is the amount of matter in an object.

Matter can be measured in weight.

A balance is a scale used to measure how much an object weighs.

The standard unit used to weigh matter is the gram.

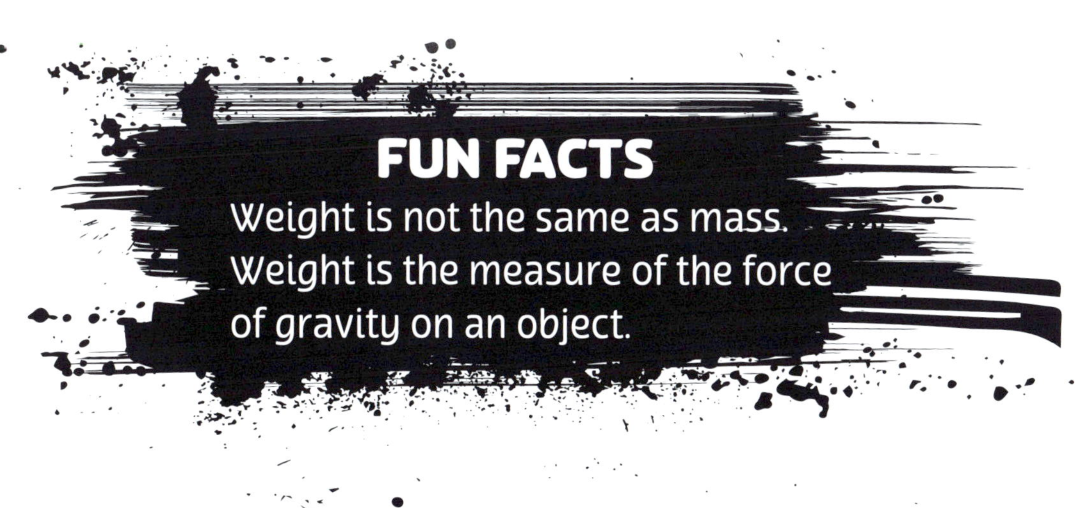

FUN FACTS

Weight is not the same as mass. Weight is the measure of the force of gravity on an object.

25
20
15
25
20
15
10
5

VOLUME

Volume is how much space an object takes up.

It is measured in different ways.

Water **displacement** is used to measure the volume of solid objects.

A **graduated cylinder** measures the volume of liquids.

FUN FACTS

People use an understanding of volume every day for jobs such as making a recipe, putting gasoline in a car, or selecting different sizes of soda bottles.

MAGNETISM

Some objects are magnetic.

They contain certain metals such as iron.

Magnets are used to determine if an object is magnetic.

Items that do not attract magnets are nonmagnetic.

FUN FACTS

Magnets are used to power speakers in earphones and televisions. They are used in computers to store information.

TEMPERATURE

Temperature is the amount of heat energy in an object.

It is how hot or cold something is.

A thermometer is the tool used to measure temperature.

Temperature can be measured in degrees Celsius or Fahrenheit.

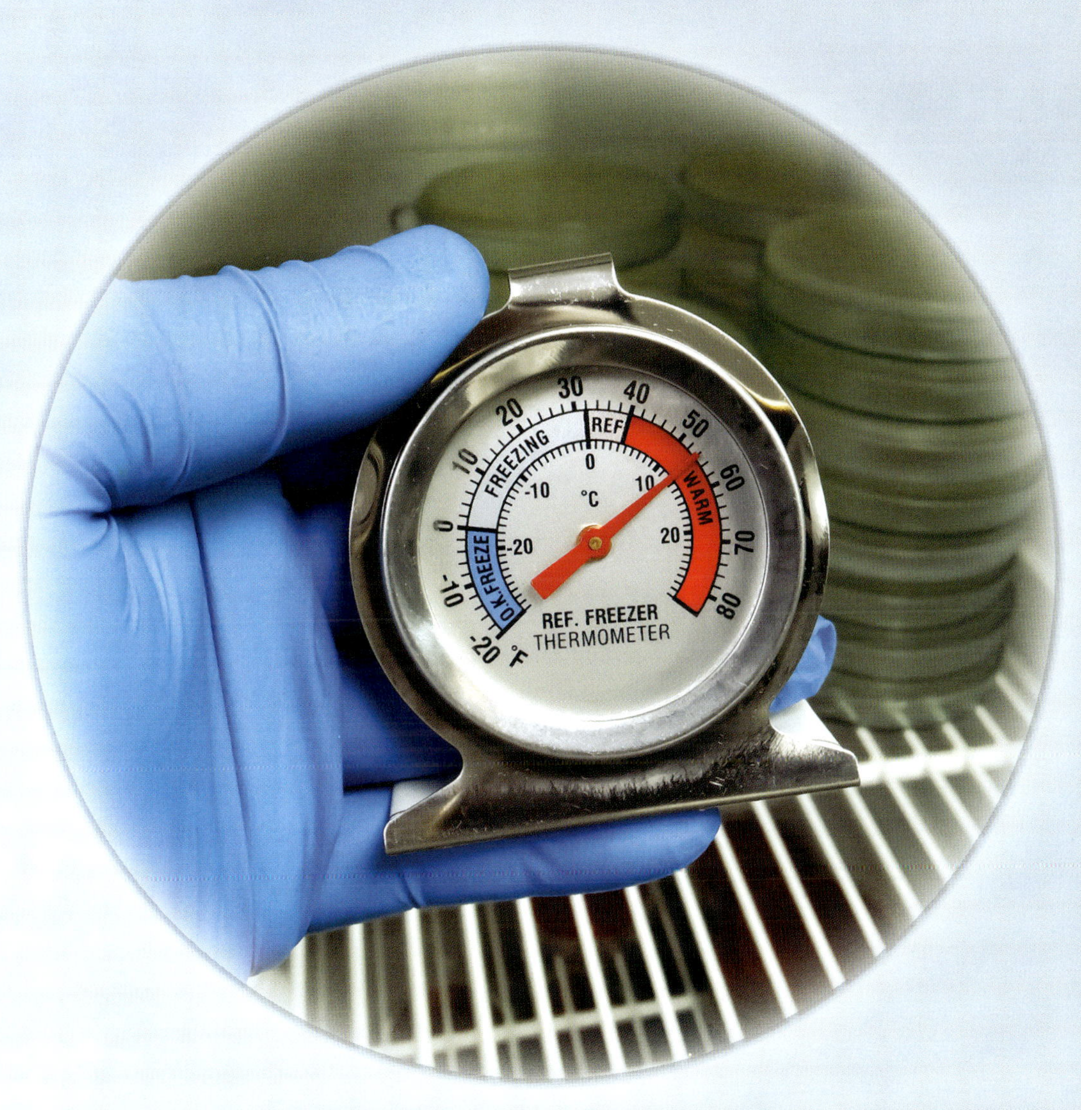
REF. FREEZER
THERMOMETER
FREEZING
REF
WARM
O.K. FREEZE
°C
°F

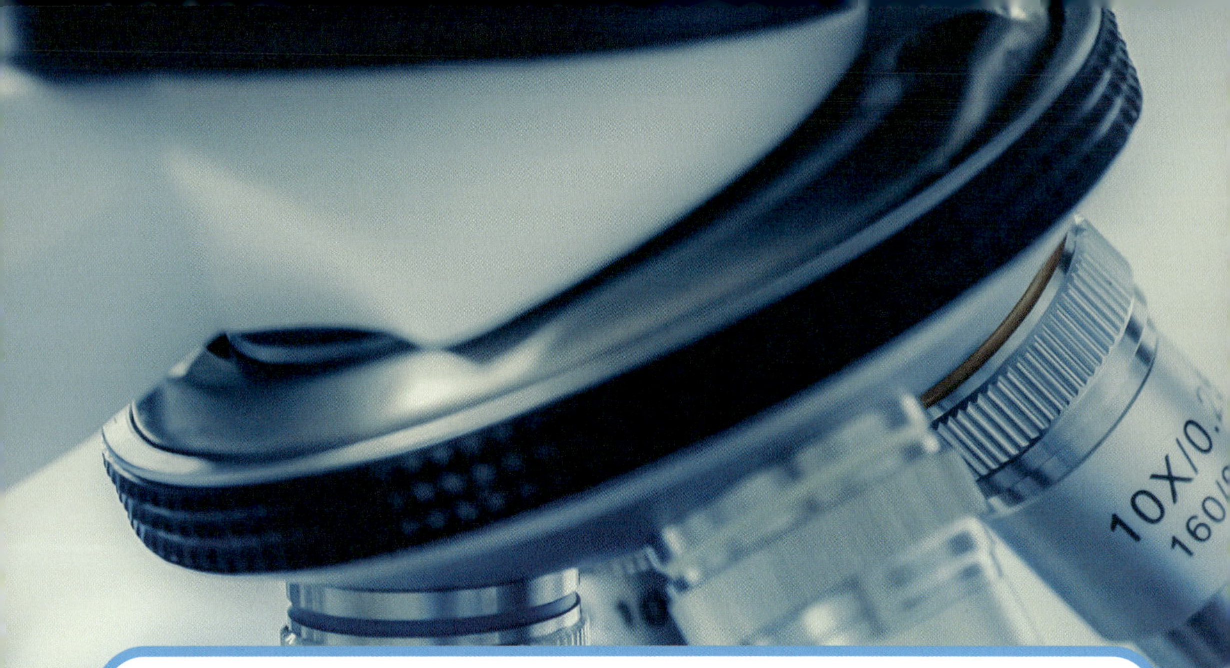

STATES OF MATTER

Matter that is solid has its own shape.

Matter that is liquid takes the shape of its container.

Matter that is a gas does not have its own shape and will fill up any available space.

These three states of matter have observable properties.

States of Matter

Solid

Liquid

Gas

FUN FACTS

Matter is made of atoms. These particles are so tiny that you need a special microscope to see them.

OBSERVABLE PROPERTIES

Observable properties are properties that cannot be measured.

They are observed using your five senses.

Shape, color, and smell are examples of observable properties.

Texture is another example. It describes how rough or smooth an object is.

Materials engineers make and test materials.

These new materials are used to make different products like golf clubs and wings on airplanes.

These scientists also create new ways to use materials that already exist.

INVESTIGATE: CANDY STORE

Materials:

- 1-3 types of mini chocolate candy
- 1-3 types of mini fruit **flavored** candy
- 1-3 types of gummy candy
- 1-2 types of hard candy
- Small cups or bowls
- Paper and pencil

Procedure:

(1) On the paper, create a chart with nine boxes.

(2) Open the candy and place in the small cups or bowls.

(3) Sort the candy into nine groups based on a single property, such as color, texture, flavor, hardness/softness, and shape. Some candies will fit into more than one group.

(4) Place each group in a box on the paper and label it with the property.

THE SCIENTIFIC METHOD

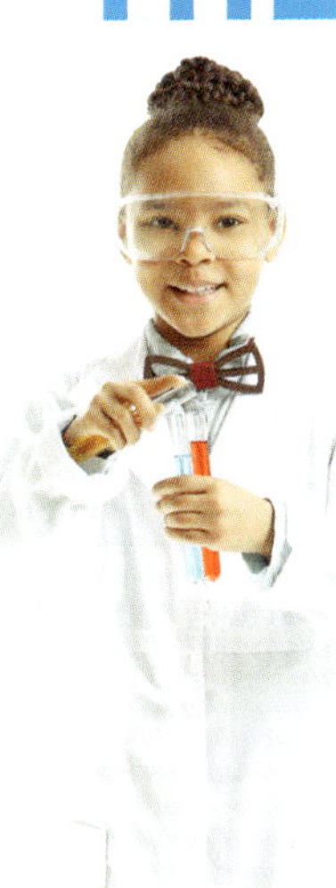

- Ask a question.
- Gather information and observe.
- Make a hypothesis or guess the answer.
- Experiment and test your hypothesis, or guess.
- Analyze your test results.
- Modify your hypothesis, if necessary.
- Make a conclusion.

SCIENTIST SPOTLIGHT

Patricia Billings invented Geobond®. This material used in buildings cannot be destroyed and is fireproof and **nontoxic**. As a sculptor, she wanted to make a plaster that wouldn't shatter and break. After many years, she was successful and made plaster that was nearly indestructible. She also learned that this material did not become hot very easily. Her invention Geobond® changed the building industry.

GLOSSARY

displacement (dis·PLAYS·muhnt): the volume of water that is displaced (or forced out) when an object is submerged

flavored (FLAY·vuhrd): made to have a particular taste

graduated cylinder (GRA·juh·way·tuhd SI·luhn·duhr): a tall narrow container with a volume scale used by scientists to measure liquids

measure (ME·zhuhr): to use a tool to determine size, amount, degree, or another property

nontoxic (nahn·TAHK·sik): not poisonous or harmful

observed (uhb·ZURVD): noticed and registered as being significant to use as data

properties (PRO·puhr·teez): attributes, qualities, or characteristics of something

texture (TEKS·chuhr): the feel, appearance, or consistency of a surface

INDEX

AFTER READING QUESTIONS

1. What is mass?
2. What is temperature?
3. What tool can be used to measure the volume of a liquid?

About the Author

Christina Earley lives in South Florida with her husband, son, and dog. Her favorite subject in school was science. She enjoys learning the science behind the world around her by asking questions such as how do roller coasters work. She loves mint chocolate chip ice cream and mermaids.

Written by: Christina Earley
Design by: Kathy Walsh
Editor: Kim Thompson

Library of Congress PCN Data
Properties of Matter / Christina Earley
Physical Science
ISBN 978-1-63897-117-7 (hard cover)
ISBN 978-1-63897-203-7 (paperback)
ISBN 978-1-63897-289-1 (EPUB)
ISBN 978-1-63897-375-1 (eBook)
Library of Congress Control Number: 2021945236

Printed in the United States of America.

Photographs/Shutterstock: Cover ©Open Studio, ©totojang1977, ©Viktoriia Debopre: Cover, Pg 1, 3, 22, 23 ©Everyonephoto Studio: Pg 4, 5 ©Stone36: Pg 6, 7©cigdem: Pg 7 ©Andrey_Popov: Pg 8, 9 ©TSViPhoto: Pg 10, 11 ©tkyszk: Pg 10 ©sirikorn thamniyom: Pg 12, 13 ©Dmitry Naumov: Pg 13 © Arif biswas: Pg 14, 15 ©Chokniti Khongchum: Pg 15 ©By Nostagrams: Pg 16, 17 ©DimaBerlin: Pg 18, 19 ©Stock-Asso: Pg 20, 21 ©santima studio: Pg 21©@ Wiki, ©Pixel-Shot

Seahorse Publishing Company
www.seahorsepub.com

Published in the United States
Seahorse Publishing
PO Box 771325
Coral Springs, FL 33077